Start and Stop

by Lola M. Schaefer

Consulting Editor: Gail Saunders-Smith, Ph.D.

Consultant: P. W. Hammer, Ph.D., Acting Manager
of Education, American Institute of Physics

Pebble Books

an imprint of Capstone Press
Mankato, Minnesota

Pebble Books are published by Capstone Press
151 Good Counsel Drive, P.O. Box 669, Mankato, Minnesota 56002
http://www.capstone-press.com

2 3 4 5 6 05 04 03 02 01

Library of Congress Cataloging-in-Publication Data
Schaefer, Lola M., 1950–
    Start and stop/by Lola M. Schaefer.
    p. cm.—(The way things move)
    Includes bibliographical references and index.
    Summary: Simple text and photographs describe and illustrate starting
and stopping.
    ISBN 0-7368-0397-1
    1. Inertia (Mechanics)—Juvenile literature. [1. Inertia (Mechanics)] I. Title.
II. Series.
QC133.5.S335   2000
531'.1—DC21   99-14035
                                                            CIP

Note to Parents and Teachers

The series The Way Things Move supports national science standards for units on understanding motion and the principles that explain it. The series also shows that things move in different ways. This book describes and illustrates starting and stopping. The photographs support early readers in understanding the text. The repetition of words and phrases helps early readers learn new words. This book also introduces early readers to subject-specific vocabulary words, which are defined in the Words to Know section. Early readers may need assistance to read some words and to use the Table of Contents, Words to Know, Read More, Internet Sites, and Index/Word List sections of the book.

Table of Contents

Start means to begin movement.

A push starts a sled.

8

A twist starts a top.

Moving air starts
a pinwheel.

Stop means
to end movement.

Brakes stop a bike.

16

A target stops an arrow.

A pull stops a horse.

A throw starts a ball.

A catch stops a ball.

Words to Know

brake—a tool that slows down or stops a vehicle; people use brakes in vehicles such as bikes, cars, and sleds.

movement—the act of changing position from place to place

pinwheel—a toy wheel that spins in the wind; people often make pinwheels with colored paper or plastic attached to a stick.

target—a mark, circle, or object at which to aim or shoot

top—a spinning toy

twist—a spin or a quick twirl using a thumb and a finger; a twist starts a top.

Read More

Challoner, Jack. *Fast and Slow.* Start-up Science. Austin, Texas: Raintree Steck-Vaughn, 1997.

Marshall, John. *Go and Stop.* Energy and Action. Vero Beach, Fla.: Rourke, 1995.

Wheeler, Jill C. *Move It!: A Book about Motion.* Kid Physics. Minneapolis: Abdo & Daughters, 1996.

Internet Sites

Eggs at Rest Stay at Rest
http://www.tryscience.org/experiments/
experiments_begin.html?eggs

Nyelabs Home Demos
http://nyelabs.kcts.org/homedemos

Index/Word List

Word Count: 49
Early-Intervention Level: 7

Editorial Credits
Martha E. H. Rustad, editor; Timothy Halldin, cover designer; Heidi Schoof, photo researcher

Photo Credits
David F. Clobes, 6
International Stock/Johnny Stockshooter, 18
Kate Boykin, 20 (top and bottom)
Leslie O'Shaughnessy, 1
Marilyn Moseley LaMantia, cover
Photophile/Bachmann, 4
Richard Hamilton Smith, 8
Unicorn Stock Photos/Nancy Ferguson, 10; Dick Young, 12; Scott Liles, 16
Visuals Unlimited/Richard C. Johnson, 14

24